Economics Efficiency of Maize Production in Gudeya Bila District, Ethiopia

Tolesa Tesema

GRIN

Bibliographic information published by the German National Library:

The German National Library lists this publication in the National Bibliography; detailed bibliographic data are available on the Internet at http://dnb.dnb.de.

ISBN: 9783346571496
This book is also available as an ebook.

Print and binding: Books on Demand GmbH, Norderstedt, Germany
Printed on acid-free paper from responsible sources.

GRIN web shop: https://www.grin.com/document/1162275

Economics efficiency of maize production in Gudeya bila district, Oromia national regional state Western parts of Ethiopia.

Tolesa Tesema: lecturer at department of Agricultural Economics, Wollega University, Ethiopia

Abstract

Farming is the ways of life for smallholder farmers of Ethiopian which is characterized with low productivity. Thus poor are failing to achieve rapid economic growth due to food insecurity. Hence the best possible means of achieving economic development is through increasing sustainable allocation resource efficiency. This study employed stochastic frointer approach to assess the levels of efficiency and Tobit model to assess the determinants of efficiency by collecting data from 154 sample household in Gudeya Bila district, western parts of Ethiopia. The result of Tobit model shows that loss due to wild animals had negative impact; slope of land and off farm income had a positive impact on technical efficiency of farmers. Concerning allocative efficiency off farm income was found to have negative effect and use of mobile was found to have positive effect. Regarding to determinants of economic efficiency slope of land has positive influence and loss due to wild animals and off farm income have negative influence. As assessed from the finding of the study farmers in the study area kills the wild animals such as monkey, pigs and apes due to their negative impact. Thus the agricultural policies and strategies of Ethiopian governments should be directed towards providing tourism, construction of soil bund, tree planting, planting grass, fencing and use of natural fertilizer to keep the fertility of steep slope and supporting the farmers by providing network facilities for mobile usage that boost maize productive efficiency of farmers.

Key words: Maize production, inefficient farmers, stochastic frontier and Tobit model

Table of Contents

1. Introduction

Farming techniques in Ethiopian have been changed little over the centuries yielding low outputs and making farmers vulnerable to the effects of unpredictable weather patterns (ATA, 2015; Tefaye and Beshir, 2014). More ever agriculture is stagnant due to inefficiency in farm production, ever rising population growth rate that was not able to meet available food demand for consumption (Nandeeswara and Bealutukela, 2015).So exposure of returning to poverty and food insecurity remains high, particularly for rural livelihoods dependent on rained agriculture (World Bank, 2016).Thus the main ways in bringing down poverty, achieving food security, sustainable development to increase standard of living for poor rural community is improvements of agricultural productivity . The major food crops are produced in almost all regions of the country has variation in volume of production across the regions attributed to the extent of area devoted to each crop type, weather change and a shift in preference for the crops grown (CSA, 2019).Regarding to area cultivation over 90% of farmers are smallholders cultivating one hectare or less of land even farming techniques have changed little over the centuries, yielding low outputs and making farmers vulnerable to the effects of unpredictable weather patterns (ATA, 2018).Even if maize is an important food crop in Gudeya Bila districts western parts of Ethiopia and the efforts made so far to generate and disseminate improved technologies, its productivity remains below its potential, Hence the community may not able educate their children, no health improvements, no life-sustenance and presence of social conflicts due to food insecurity and increase in price of commodity that results severe underdevelopments. The measurement of efficiency difference is due to heterogeneity of conditions and the diversity of environments in which farmers operate does not have to be unique (Mechri, *et al.*, 2017).Moreover it is not sufficient to meet national requirement of production by only yield-enhancing agricultural technologies in the country, so country is still under food aid from western country(Tura *et al.*, 2010 ;Sorsie *et al.*, 2015 and Kitila and Alemu, 2014). In the study area productivity of agriculture is very low that results food insecurity and no life substance. So it is important to update information and knowledge's gab on efficient allocation of resource since social developments is dynamic and geographical difference. Thus this study efficient allocation of resource and identifying the determinants of efficiency by adding new variables such as loose due to wild animals and uses of mobile on information transmission between the farmers and government agency.

2. Methods

2.1. Description of the study area

This study was carried out in western parts of Ethiopia in low lands of Gudeya Bila districts. The farming system in the district is mainly mixed crop-livestock production in which livestock provides source of manure for crop and the residual from crop output was used as source of feeding for livestock. Agriculture is mainly characterized by rain-fed production system which is used for livelihood sustenance. Maize is one of the major cereals grown in the study area which farming is mostly with limited oxen and no mechanized faring system. The district is bordered by Jima Ganeti and BakoTibe districts in east, Guto Gida and Sibu Sire districts in west, Abe Dongoro district in north and Gobbuu Sayyoo district in south. It is located at 104 from the zonal capital and 274km from Addis Ababa, capital of Ethiopia to west. It lies between 370 01' 28"N latitude and 90 17'23" S longitudes. Altitude ranges between 500 to 3500 meters above sea level (GBWOANR, 2020)

2.2. Sampling techniques and questionaries' design

Multi stage sampling techniques was used in this study. In the frist stage Gudeya Bila was selected from western parts of Ethiopia since maize was dominant crop in the district and the researcher is convenient to the study area. In the second stage three Kebeles of namely Zangi, Gute Chancho and Abay Dale were purposively selected from the district since they are low land of the district and maize is the dominant food crop grown in the study area and the option for livelihood sustenance and food security of major crop in the study area. In the third stage 60, 24 and 70 sample household was selected by simple random sampling from three kebeles.finally 154 households that produce maize were selected by probability proportional to sample size in the low lands of districts. This research employed cross-sectional survey using structured questionnaire which was distributed to the sample household to collect amounts of output produced and amounts of inputs used, constraints of maize production and the socioeconomic, demographic, institutional and farm-specific characteristics, input price, cost of input incurred during maize production period such as amount and cost of labour used, rental cost of land, cost of oxen used, and cost seed and fertilizer.

2.3. Model specification for efficiency and determinants of efficiency

Stochastic frontier model was employed to estimate parameters of production function and level of efficiency. This is because of the fact that this technique accounts for measuring inefficiency factors and technical errors occurring during measurement and observation (Coelli, 1998). So as to take into account effects of these errors, stochastic frontier model was used in this study. Following Aigner *et al.* (1977) and Meeusen and van Den Broeck (1977), the stochastic frontier model is defined below was adopted for this study.

$$Y_i = f(X_i ; \beta) + v_i - u_i \qquad 1$$

Where Y_i measures the quantity of maize output of the ith farm in the Gudeya Bila districts, X_i is the vectors of inputs variables used by the farmer i^{th} in the low land such as land, labor, urea fertilizer, NPS fertilizer, oxen and seed used by sample household during 2020/2021 production season.

The β is the vector of unknown parameters. The functional specification $f(X_i ; \beta)$ is a proper production function (Cobb-Douglas). The disturbance term v_i is intended to capture the effects of the stochastic noise and it is assumed to be v_i~N $(0,\delta^2)$. The disturbance, ui, captures the technical inefficiencies.

This study adopted Sharma et al. (1999), to estimate the technical, allocative and economic efficiencies respectively adopted for this study. In this study technical efficiency levels is obtained in terms of observed output (Yi) food maize produced to the corresponding frontier output (Yi*) using the available technology.

$$TEi = \frac{Y_i}{Y^*} \qquad 2$$

The farm specific economic efficiency is obtained as the ratio of minimum observed total Production cost (C*) to actual total production cost (C).

$$EE\ i = \frac{C_*}{C} \qquad 3$$

Following Farrell (1957), the allocative efficiency (AE) index can be derived as ratio of economic efficiency (EE) to technical efficiency (TE):

$$AE = \frac{EE}{TE} \qquad 4$$

After estimating the level of technical, allocative and economic efficiency from Stochastic production frontier they were regressed using a censored Tobit model on farm specific independent variables that was shows disparity in efficiency across farms. Tobit regression (Gujarati, 2004) is specified as:

$$E^* = \delta o + \delta_m Z_{im} + \mu \qquad 5$$

$$v / z \approx normal(0, \delta^2)$$

Where E^*-latent variable representing the technical, allocative and economic efficiency scores of maize producer in the study area.

$\delta -$ a vector of unknown parameter to be estimated.

Z_{im} $-$ a vectors of explanatory variables m(m=1,2,...n)for maize producres such as

X1= loss due to wild animals such as ape, pig and monkey during pre-harvest stage

X2= Education levels of household in year of schooling

x3 =Family size in man equivalent

x4=Sex of household in dummy (1 if male zero if female heeded household

x5= Farm size in hectare

x6=Livestock holding in tropical livestock unit

X7=Livestock in tropical livestock unit

X8=slope of land measured in dummy (1 if flat and zero if steep)

X9=Fertility which is either fertile or infertile

X10= distance to plot in minute

X11= Land conservation which is one for plot which is conserved zero otherwise

X12=Ploughing frequency in number

X13= Distance to the markets market in minute

X14=Extension contacts in frequency of contacts during production season

X15=Use of radio measured dummy (one if farmers use radio zero if not)

X16=Uses of mobile measured as dummy (one 1 if farmers used mobile zero if not)

X17=Credit uses measured in dummy (one if farmers received credit zero if not)

X18=Off farm income dummy (one if farmers engaged in off farm activities zero if not)

$\mu -$ Error term that is independently and normally distributed with zero mean and variance δ^2

Denoting E_i as observed variables,

$$E_i = \begin{Bmatrix} 1 & if & E_i^* >= 1 \\ 0 & if & E_i^* <= 0 \end{Bmatrix} \qquad 6$$

Adopting following McDonald and Moffitt (1980 from the likelihood function decomposition of marginal effects two-limit Tobit model are :

The unconditional expected value of the dependent variable

$$\frac{\partial E(y)}{\partial X_j} = [\varphi(Z_u) - \varphi(Z_L)].\frac{\partial E(y^*)}{\partial X_j} + \frac{\partial[\varphi(Z_u) - \varphi(Z_L)]}{\partial X_j} + \frac{\partial[1 - \varphi(Z_u)]}{\partial X_j} \dots\dots\dots (7)$$

The expected value of the dependent variable conditional upon being between the limits

$$\frac{\partial E(Y^*)}{\partial X_j} = \beta m.\left[1 + \frac{\{Z_L\phi(Z_L) - Z_u\phi(Z_u)\}}{\{\varphi(Z_u) - \varphi(Z_L)\}}\right] - \left[\frac{\{\phi(Z_L) - \phi(Z_u)\}^2}{\{\varphi(Z_u) - \varphi(Z_L)\}^2}\right] \dots\dots \quad \dots\dots\dots\dots (8)$$

The probability of being between the limits

$$\frac{\partial[\varphi(Z_u) - \varphi(Z_L)]}{\partial X_j} = \frac{\beta_m}{\sigma}[\phi(Z_L) - \phi(Z_u)] \dots\dots\dots\dots\dots\dots\dots\dots\dots\dots\dots\dots\dots\dots\dots. (9)$$

Where, $\varphi(.)$ = the cumulative normal distribution,

$\phi(.)$ = the normal density function

ZL $= -\frac{X_i'\beta}{\sigma}$ and Zu $=\frac{(1-X_i'\beta)}{1}$ are standardized variables that came from the likelihood function given the limits of y* and σ = standard deviation of the model.

3. Results

3.1. Estimation maximum likelihood estimate

The maximum likelihood estimates of the parameters of the frontier production functions are presented in table 1.From the total of five variables considered in the stochastic production function, three variables (seed, land and labor) had a significant effect that makes different in productivities of maize. As a result, 1% increase in amount of land will result in 0.34% increase in maize production, keeping other factors constant. The scale coefficient was calculated to be 1.031 indicating increasing returns to scale (Table 2).This implies that there is potential for maize producers to continue to swell their production because they are in the stage I of the production surface, where resource use and production is believed to be underutilized.

Table 1.Estimation of the Cobb-Douglas frontier production function

Loutput	Coef.	Std. Err.	z	P>z
Lan seed	0.21758998*	0.1175298	1.85	0.064
Lan land	0.34213118***	0.1064939	3.21	0.001
Lan NPS	0.07337987	0.079243	0.93	0.354
Lan UREA	0.00434708	0.1004711	0.04	0.965
Lan oxen	0.28640016***	0.1039812	2.75	0.006
Lan labor	0.10739182	0.0839132	1.28	0.201
Cons	1.419079	0.4895973	2.90	0.004
Sigma square	0.4997371	0.0700897		0.3623639
Lambda	4.929533	0.0724496		4.787535

Note:*, ** and *** refers to10%, 5% and 1% significance level, respectively.

Source: Stochastic frontier model result (2021)

The dual cost function derived analytically from the stochastic production function is given as follows basis for computing allocative and economic efficiency

$$lCmi = 4.27 + 0.25\omega_1 + 0.076\omega_2 + 0.0.34\omega_3 - 0..14\omega_4 + 0..17\omega_5 - 0.01\omega_6 + 0.14Y*$$

Where $lCmi$ is cost of producing maize; ω_1 refers to the price of seed, ω_2 is cost of land; ω_3 is cost of NPS; ω_4 is cost of UREA; ω_5 is cost of oxen; ω_6 is cost of labor Y^* is output adjusted for any statistical noise; i^{th} refers to the ith sample household.

3.2. Technical, allocative and economic efficiency score

The result of frontier model revealed that, farmers in the study area were relatively good in allocative efficiency than in technical and economic efficiency of maize production. The mean technical efficiency level was 63.86% varies from 4.04 to 94.44 percent shows that maize producing farmers have an opportunity to efficiently utilize resources and hence they could increase the current maize output by 36.14% using the existing technology. The means allocative efficiency of farmers in the study area was 68.41% and ranges from 37.73% to 88.5% indicating that on average, maize producer farmers can save 31.59% of their current cost of inputs if resources are efficiently utilized. As illustrated in the above table 4 the mean economic efficiency level of sample households was 43.65% with minimum and maximum efficiency scores of 1.78% and 79.79% respectively. That is the producer with an average economic efficiency level could reduce current average cost of production by 56.35% to achieve the potential minimum cost level without reducing output levels (Table 4). Technical efficiency result obtained from this analysis is above result of (Gosa and Jema, 2016). Allocative efficiency is below it.

Table 2.Technical, allocative and economic efficiency score

Variable	Observation	Mean	Std. Dev.	Min	Max
TE	154	0.6385534	0.1902531	0.0404119	0.9445699
AE	154	0.6841796	0.1211773	0.3773368	0.8850588
EE	154	0.436568	0.141425	0.0178012	0.7979941

Note:*, ** and *** refers to10%, 5% and 1% significance level, respectively.

Source: model result (2021)

3.3. Distribution of technical allocative and economic efficiency

Regarding to the distribution of efficiency 75.9%, 91.55% and 64.28% percent of the respondents has the technical allocative and economic efficiency score greater than 50 percent while 24.02%, 8.44%,35.71 percent's of the respondents have the technical, allocative and economic efficiency score of less than 50 percent's (Figure 2)

Figure 1.Distribution and kernel density estimates of technical allocative and economic efficiency score

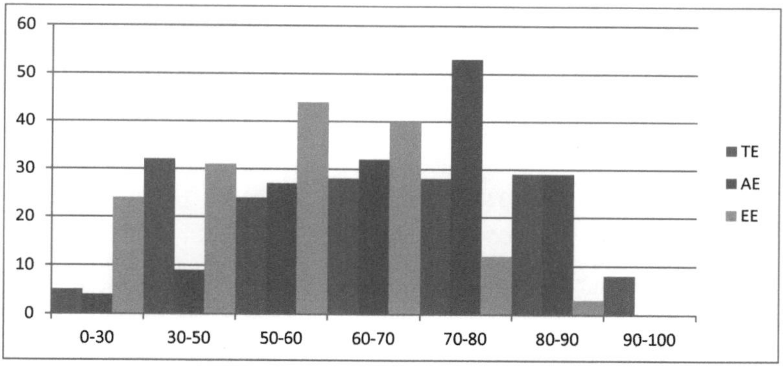

From left to right (one bar each): TE, AE, EE

Source: The model result

3.4. Determinants of Technical, allocative and economic efficiency of maize and its marginal effects

In the study area it is common that the loss due to wild animals such as ape, monkey and pigs consume the maize crop in day and night period during pre-harvest stage. So it is difficult to control for the farmers such wild animals even iff they devote more time to control thus wild animals during that time. The coefficient of maize crop loosed due to its consumed by wild animals such as ape, monkeys and pigs for technical efficiency and economic efficiency is negative and statistically significant at 5% and 10% percent significance level respectively. This is due to farmers wastes their time more on protecting wild animals from crop rather than investing their time on crop weeding of the crop that increase its yields. These results of marginal effect shows there is probability of farmers technically and economically efficient by 0.56% and 0.03%,improvements of mean technical and economic efficiency by 0.92% and 0.66% and overall increase in technical and economic efficiency by 0.56 0.67%.

Table 3.Determinants of technical efficiency and its marginal effects

TE	Coefficients	Standard error	Marginal effects		
			$\frac{\partial E(y)}{\partial X_j}$	$\frac{\partial E(y^*)}{\partial X_j}$	$\frac{\partial(\phi(Z_U - Z_L)}{\partial X_j}$
Loss due to wild animals	-0.0112588**	0.0050096	0.005647	0.0092759	0.005647
Educational levels	0.0036549	0.0032843	0.0018331	0.0030112	-0.0018331
Family size	-0.0030456	0.0055415	0.0015275	-0.0025092	0.0015275
Farm experience	-0.0040234	0.0031703	0.002018	-0.0033148	0.002018
Sex of household	0.0574569	0.0379045	0.0244649	0.0483981	-0.0244649
Farm size	0.023305	0.0246989	0.0116889	0.0192005	0.0116889
Livestock holding	-0.0031388	0.0049439	0.0015743	-0.002586	0.0015743
Slope of land	0.0525587***	0.0185236	0.0263614	0.043302	0.0263614
Fertility status	0.0405947	0.0306413	-0.0214925	0.0331732	-0.0214925
Distance to plot	0.0002111	0.0014108	-0.0001059	0.0001739	-0.0001059
Land conservation	-0.0165475	0.0300147	0.0083595	-0.0136187	0.0083595
Ploughing frequency	-0.0071782	0.0213898	0.0036003	-0.005914	0.0036003
Distance to the markets	-0.0004811	0.0006866	0.0002413	-0.0003964	0.0002413
Extension contacts	-0.000498	0.0030691	0.0002498	-0.0004103	0.0002498
Use of radio	-0.0239619	0.0314711	0.0120541	-0.0197334	0.0120541
Uses of mobile	0.0060698	0.0319237	-0.003052	0.004999	-0.003052
Credit uses	-0.0193715	0.0312802	0.0096506	-0.0159759	0.0096506
Off farm income	-0.0611683**	0.0362075	0.0264527	-0.0514254	0.0264527
cons	0.589492	0.1405304			

Note: *, **and *** significant at 10%, 5% and 1% level significance, respectively

$\frac{\partial E(y)}{\partial X_j}$ $\frac{\partial E(y^*)}{\partial X_j}$ $\frac{\partial(\phi(Z_U - Z_L)}{\partial X_j}$ Refers to total change, expected change and changes of probability

Source: Tobit model results (2021)

The result also indicated that use of mobile had a positive sign and statistically significant effect on allocative efficiency level at 10% level of significance. This suggests that on average households with use mobile exhibit higher levels of allocative efficiency. This due to the reason that use of mobiles allows a household to enhance knowing the production method and information since Ethiopian telecom is disseminating production, price and rain fall information to the farms through Telephone. This result is consistence with (Sisay et al.2015)The results of marginal effect indicates that the probability of farmers allocatively efficient increase by 2.55%,mean allocative efficiency increase by 3.01% and overall increase in probability and mean efficiency increase by 3.59%.

Table 4.Tobit model estimates for determinant of allocative efficiency and its marginal effects

Allocative efficiency	Coefficients	Standard error	Marginal effects		
			$\frac{\partial E(y)}{\partial X_j}$	$\frac{\partial E(y^*)}{\partial X_j}$	$\frac{\partial(\phi(Z_U - Z_L)}{\partial X_j}$
Loss due to wild animals	0.00035055	0.0032408	0.0003352	0.000282	-0.0002294
Educational levels	0.00282156	0.0021263	0.0026977	0.0022696	-0.0018462
Family size	-0.00084345	0.0035892	-0.0008064	-0.0006785	0.0005519
Farm experience	-0.00292775	0.002055	-0.0027993	-0.0023551	0.0019157
Sex of household	0.02697432	0.0245365	0.0259049	0.0219488	0-.0149036
Farm size	-0.01248	0.0159613	-0.0119323	-0.0100388	0.0081659
TLU	0.00448694	0.0031909	0.00429	0.0036093	-0.0029359
SLOPE	0.00080154	0.0119793	0.0007664	0.0006448	-0.0005245
FERTLTY	0.00755362	0.0198128	0.0072172	0.0060658	-0.0050407
DIPLOT	-0.00008074	0.0009148	-0.0000772	-0.0000649	0.0000528
Land conservation	0.01808851	0.0194179	0.0172996	0.0145623	-0.0116785
Ploughing frequency	0.01298822	0.0138426	0.0124182	0.0104476	-0.0084985
Distance to the markets	-0.00030382	0.0004448	-0.0002905	-0.0002444	0.0001988
Extension contacts	-0.00184633	0.0019856	-0.0017653	-0.0014852	0.0012081
Use of radio	-0.02572104	0.0203828	-0.0245797	-0.0206679	0.0169398
Uses of mobile	0.03767349*	0.0206359	0.0359526	0.030177	0.0255745
Credit uses	0.00901042	0.0202362	0.0086125	0.0072428	-0.0059414
Off farm income	-0.05427372**	0.0234452	-0.0522163	-0.0443886	0.0262476
cons	0.58158188***	0.0909639			

Note: *, **and *** significant at 10%, 5% and 1% level significance, respectively

$\frac{\partial E(y)}{\partial X_j}$ $\frac{\partial E(y^*)}{\partial X_j}$ $\frac{\partial(\phi(Z_U - Z_L)}{\partial X_j}$ Refers to total change, expected change and changes of probability

Source: Tobit model results (2021)

The sign of slope of land on was positive effect on technical and economic efficiency at 1% level of significance. This implying farmers who are plotting the maize on steep slow are lee efficient than farmers who are plotting on flat slops. This could be because the flat slops are not susceptible to the soil erosion and the fertilizers are not displaced from the sowed place This result was in line with the findings of (Mustefa, 2017; Kitila and Alemu, 2014).More ever the results of marginal effect of Tobit model shows that the probability of farmers technically and economically efficient by due to slope of land is flat is 2.63% and mean technical and economic efficiency due to slope of land is flat is 4.33% and 3.71% and the overall increase in probability and means efficiency due to slope of land is flat is 2.63% and 3.81%

Table 5.Tobit model estimates for determinant of economic efficiency and it marginal effects different efficiency

Economic efficiency	Coefficients	Standard error	Marginal effects		
			$\frac{\partial E(y)}{\partial X_j}$	$\frac{\partial E(y^*)}{\partial X_j}$	$\frac{\partial(\phi(Z_U - Z_L)}{\partial X_j}$
Loose due to wild animals	-0.00681933*	0.0036471	-0.0067977	-0.0066199	0.0003124
Educational levels	0.0046077	0.0024256	0.0045931	0.0004473	-0.0002111
Family size	-0.00256184	0.0040359	-0.0025537	-0.0024869	0.0001174
Farm experience	-0.00483249**	0.0023373	-0.0048171	-0.0046912	0.0002214
Sex of household	0.05312193	0.097616	0.0529699	0.051671	-0.0011557
Farm size	0.00260418	0.0179501	0.0025959	0.002528	0.0001193
Livestock holding	0.00181602	0.0035974	0.0018102	0.0017629	-0.0000832
Slope of land	0.03829706***	0.0134816	0.0381753	0.0371771	0.0017543
Fertility status	0.02373793	0.0222869	0.0236581	0.0230183	-0.0012202
Distance to plot	-0.00057302	0.0010257	-0.0005712	-0.0005563	0.0000262
Land conservation	-0.00369407	0.0218472	-0.0036823	-0.0035859	0.0001702
Ploughing frequency	0.00354498	0.0155748	0.0035337	0.0034413	-0.0001624
Distance to the markets market	-0.00025624	0.0005016	-0.0002554	-0.0002487	0.0000117
Extension contacts	-0.00228599	0.0022352	-0.0022787	-0.0022191	0.0001047
Use of radio	-0.02879418	0.0229347	-0.0287007	-0.027941	0.0013445
Uses of mobile	0.01658778	0.0232058	0.0165342	0.0160979	-0.0007817
Credit uses	-0.00997011	0.0227554	-0.0099385	-0.0096792	0.0004506
Off farm income	-0.0566933**	0.0263701	-0.056528	-0.0551269	0.0013512
_cons	.33438097***	0.1025188			

Note: *, **and *** significant at 10%, 5% and 1% level significance, respectively

$\frac{\partial E(y)}{\partial X_j}$ $\frac{\partial E(y^*)}{\partial X_j}$ $\frac{\partial(\phi(Z_U - Z_L)}{\partial X_j}$ Refers to total change, expected change and changes of probability

Source: Tobit model results (2021)

In this study the coefficient of off farm activity was negative statistically significant at 5%, level of significance effect with respect to technical, allocative and economic efficiency all together respectively. Off/non-farm activities may affect the efficiency negatively for the reason that the farmers who are engaging in the off farm activities may shifts to the off farm such us petty trade, hand craft and carpentry. The result is in line with the findings of (Kifle 2017; Gizachew, 2018).As computed from the Tobit model the probability of farmers technically, allocatively and economically efficiency duet to of farm income earning was 2.64%,2.62% and 0.13%,the mean technical, allocative and economic efficiency increase due to off farm earning was 5.14%,4.43% and 5.51% and the overall increase in probability and mean efficiency was 2.64%,5.22% and 5.65%.

The experience of the household influenced economic efficiency negatively. This suggested that younger farmers were more efficient than their older counterparts. Older farmers are reluctant with the change of new technology and advancements in techniques of production. The result of marginal effect after Tobit model shows the probability of farmers technically efficiency due to increase in experience of household was 0.02%,the means economic efficiency increase by 0.46% due to increase in experience of household and the overall increase in probability and mean economic efficiency increase by 0.48%. Sisay Debebe (2015) found the similar result.

4. Conclusion and policy implication

Food production in developing country can be increased throughout improving the efficiency and solving the problem related to the production efficiency since use of improved technologies is expensive because the farmers are subsistence farmers. In the study area since monkey, pigs and apes affects the product of maize. These have two impacts in one direction there is loss of crop output and in other direction there is extinction of those species. Thus to protect those wild animals from extinction and loss of agricultural product the agricultural policies and strategies of thus providing controlling mechanism of those animals through tourism is very essential. The government of Ethiopia has to provide educational opportunities for peoples who are in the remote area on how they use mobile phone and provision of mobile phone at low cost for agricultural information transformation as well as the construction of network facility to the rural communities. Additionally improving land productivity status by applying new soil conservation practices on their farm through improved sustainable land management practices especially planting grass and soil bund is encouraged specially on steep slope since it is susceptible to soil erosion. Lastly the further research have to done on impacts of environmental degradation of agricultural efficiency, comparative analysis of agricultural efficiency in different agro ecology of the study area in particular and Ethiopia in general.

Abbreviation

ATA=Agricultural Transformation Agency ,CSA: Central Atatistical agency,TE: Technical efficiency,AE=allocative efficiency,EE=Economic efficiency,GBWOANR =Gudeya Bila Woreda office of Agriculture and Natural Resource

Acknowledgements: We gratefully acknowledge Gudeya bila office of agriculture for their support during data collection.

Reference

Agricultural Transformation Agency (ATA). 2018. The agriculture knowledge, learning, documentation and policy project. The akldp project 2014-2018.adis Ababa, Ethiopia

Agricultural Transformation Agency(ATA).2015. Woreda-Level Crop Production Rankings in Ethiopia: A Pooled Data Approach. International Food Policy Research Institute (IFPRI) Agricultural annual report of Ethiopia.

Aigner, D., Lovell, C.K. and Schmidt, P., 1977. Formulation and estimation of stochastic frontier production function models. *Journal of econometrics*, *6*(1), pp.21-37.

Coelli, T., D.S.P. Rao and G.E. Battese,1998. An introduction to efficiency effects in stochastic frontier function for panel data. Empirical Economics, **20:** 325-332.

CSA(Central Statistical Agency).2019. The federal democratic republic of Ethiopia central statistical agency agricultural sample survey 2019/20.volume I report on area and production of major crops. Addis Ababa, Ethiopia

GBWOANR (Gudeya Bila Woreda office of Agriculture and Natural Resource), 2020. Gudeya Bila Woreda Office of Agriculture and Natural Resource, the annual report. Nekemte, Ethiopia.

Gizachew, F. 2018 .Allocative Efficiency of Smallholder Wheat Producers in Damot, *Journal of food science and quality management*, 72, pp. 27–35.

Gosa Alemu and Jema Haji. 2016. Economic Efficiency of Sorghum Production for Smallholder Farmers in Eastern Ethiopia: The Case of Habro District. *Journal of Economics and Sustainable Development. Vol.7,15*

Gujarati, D.2004. Basic Econometrics. McGraw-Hill Companies. Tokyo.

Kifle Degefa et al., 2017.Economic Efficiency of Smallholder Farmers in Maize Production in Bako Tibe District, Ethiopia. Vol.7, No.2.

Kitila, G. M., & Alemu, B. A. 2014. Analysis of Technical Efficiency of Small Holder Maize Growing Farmers of Horo Guduru Wollega Zone, Ethiopia: A Stochastic Frontier Approach. Science, *Technology and Arts Research Journal*, 3(3), 204-212.

Mcdonald, J. F., Moffitt. R. A. (1980). The use of Tobit Analysis. Review of Economics and Statistics. Volume 62 Issue(2), 318-321.

Mechri Aicha, Peter Lys and Franck Cachia, 2017. Productivity and Efficiency Measurement in Agriculture. Technical Report Series.

Meeusen, W. and van Den Broeck, J., 1977. Efficiency estimation from Cobb-Douglas production functions with composed error. *International economic review*, pp.435-444.

Mustefa Bati, Mulugeta Tilahun and Raja Kumar Parabathina, 2017. Economic efficiency in maize production in Ilu Ababor zone, Ethiopia. Research Journal of Agriculture and Forestry Sciences,Vol. 5(12), 1-8

Nandeeswara rao p and Bealutukela.2015. Analyzing productivity in maize production: the case of boricha woreda in sidama zone, southern Ethiopia. *International journal of recent scientific research vol. 6, issue, 10, pp. 6984-6989.*

Sharma, K.R., Leung, P. and Zalleski, H.M.1999.The technical, allocative, and economic efficiencies in swine production in Hawaii: A comparison of parametric and non-parametric approaches, Agricultural Economics, 20, 23-35.

Sisay Debebe , Jema Haji , Degye Goshu and Abdi-Khalil Edriss.2015.Technical, allocative, and economic efficiency among smallholder maize farmers in Southwestern Ethiopia: Parametric approach. *Journal of Development and Agricultural Economics, Vol. 7(8), pp. 283-292*

Sorsie Deme, Nicolette Matthews and Janus Henning.2015.Analysis of factors affecting technical efficiency of smallholder maize farmers in Ethiopia, Conference Paper.

Tefaye, W., & Beshir, H.2014. Determinants of Technical Efficiency in Maize Production: The Case of Smallholder Farmers in Dhidhessa District of Illuababora Zone, Ethiopia. *Journal of Economics and Sustainable Development*, 5(12), 274-284

World Bank, 2016. Federal democratic republic of Ethiopia priorities for ending extreme poverty and promoting shared prosperity systematic country diagnostic.